science all around me

Materials

Karen Bryant-Mole

First published in Great Britain by Heinemann Library, Halley Court, Jordan Hill, Oxford OX2 8EJ
a division of Reed Educational & Professional Publishing Ltd.

OXFORD FLORENCE PRAGUE MADRID ATHENS MELBOURNE AUCKLAND KUALA LUMPUR SINGAPORE TOKYO IBADAN NAIROBI KAMPALA JOHANNESBURG GABORONE PORTSMOUTH NH (USA) CHICAGO MEXICO CITY SAO PAULO

© BryantMole Books 1996

All rights reserved. No part of this publication may be reproduced, stored in a retrieval system, or transmitted in any form or by any means, electronic, mechanical, photocopying, recording, or otherwise without either the prior written permission of the Publishers or a licence permitting restricted copying in the United Kingdom issued by the Copyright Licensing Agency Ltd, 90 Tottenham Court Road, London W1P 9HE

Designed by Jean Wheeler
Commissioned photography by Zul Mukhida
Consultant – Hazel Grice
Printed in Hong Kong

03
10 9 8

British Library Cataloguing in Publication Data

Bryant-Mole, Karen
Materials. - (Science all around me)
1. Materials - Juvenile literature 2. Science - Juvenile literature
I. Title
620.1'1

ISBN 0 431 07827 0

A number of questions are posed in this book. They are designed to consolidate children's understanding by encouraging further exploration of the science in their everyday lives.

Words that appear in the text like this can be found in the glossary.

Acknowledgements
The Publishers would like to thank the following for permission to reproduce photographs: Chapel Studios 6, 12; Eye Ubiquitous 4, 8, 18, 20; Positive Images 16; Tony Stone Images 10 (Paul Chesley), 14 (John Lamb), 22 (Paul Chesley).

Every effort had been made to contact copyright holders of any material reproduced in this book. Any omissions will be rectified in subsequent printings if notice is given to the Publisher.

Contents

What are materials?

Some people use the word 'material' to mean cloth or fabric. But in science, material means anything that things are made from.

Everything is made from something. Lots of different materials have been used to make the things in this picture.

Material does not just mean fabric, but fabric is a type of material.

Some things are made from just one material.
Others are made from more than one material.

The bowl is made from one material.
The scissors are made from two different materials.

Is the watering can made from one material or two?

Materials all around us

Everywhere you look, you see materials.
Different materials have different names.

The toy plane in this photograph is made from wood.

The car is mostly made from metal. The truck is made from plastic.

? *What are the pages of this book made from?*

See for yourself ...

James has collected together some objects.
Can you see three things that are made from wood, three that are made from metal and three that are made from plastic?

Look around you.
What other materials can you see?

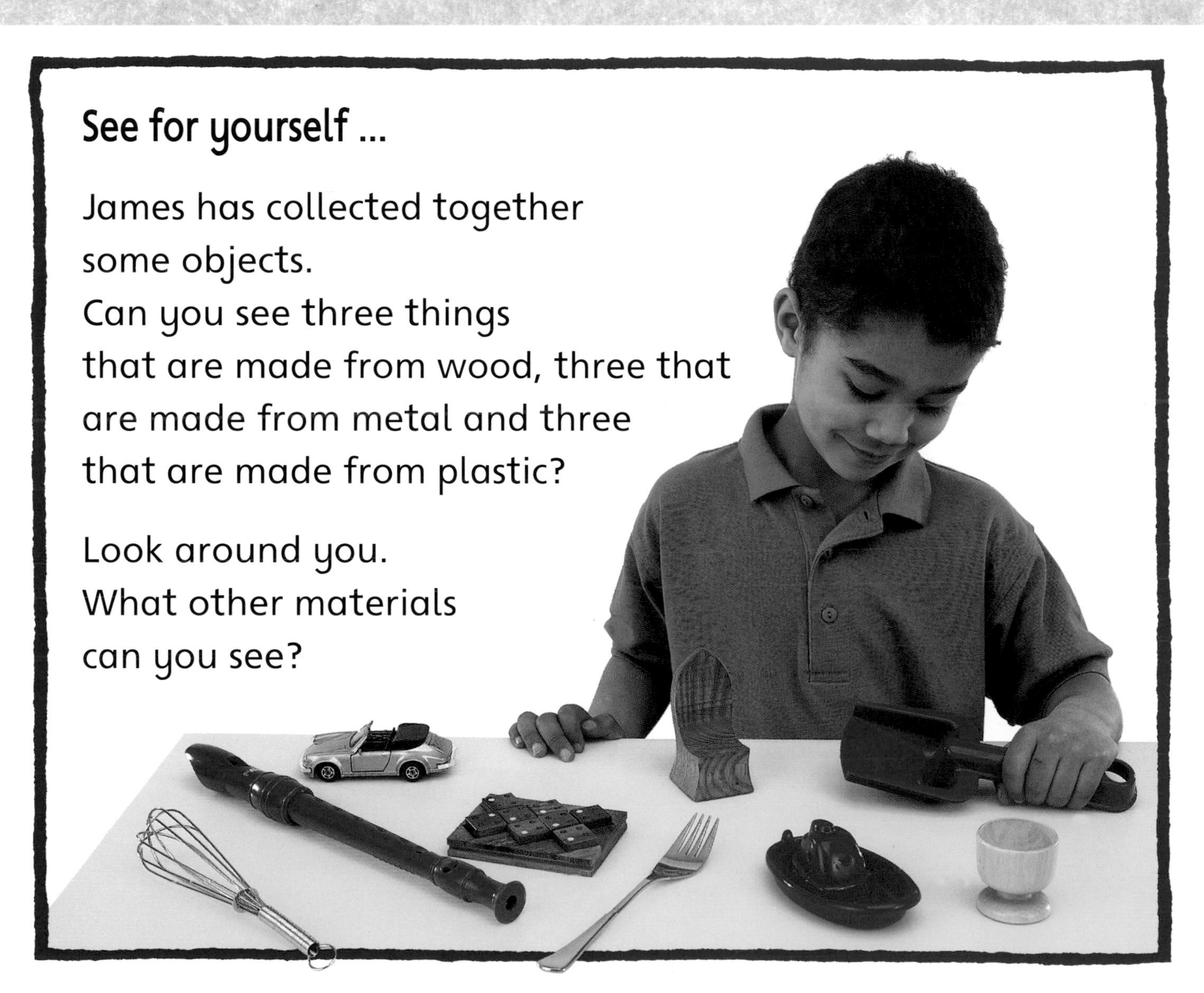

Natural materials

'Natural' means something that is part of nature. Natural materials come from animals or plants or are dug up from the ground.

Clay is a natural material. It is a special type of soil. Clay can be used to make things such as cups and bowls.

? *Can you find something that is made from clay?*

See for yourself ...

Naheed's dad helped Naheed to find some objects made from natural materials.

Naheed is trying to work out which come from animals, which come from plants and which are dug up from the ground.

Can you help him? (There are two of each.)

Synthetic materials

Synthetic materials are usually made in factories.
Plastic is a synthetic material.
It is made from chemicals that are found in oil.

Many of today's fabrics are made from synthetic **fibres**.
They are made into clothes that are easy to wash and comfortable to wear.

'Synthetic' comes from a word that means 'put together'.

See for yourself ...

Are your clothes made from synthetic fibres or natural fibres?

Wool, silk and cotton are natural fibres. They come from animals and plants.

Nylon, acrylic, viscose and any fabrics that begin with the letters 'poly' are synthetic.

Jade is looking at the labels on some clothes and sorting them into two different piles.

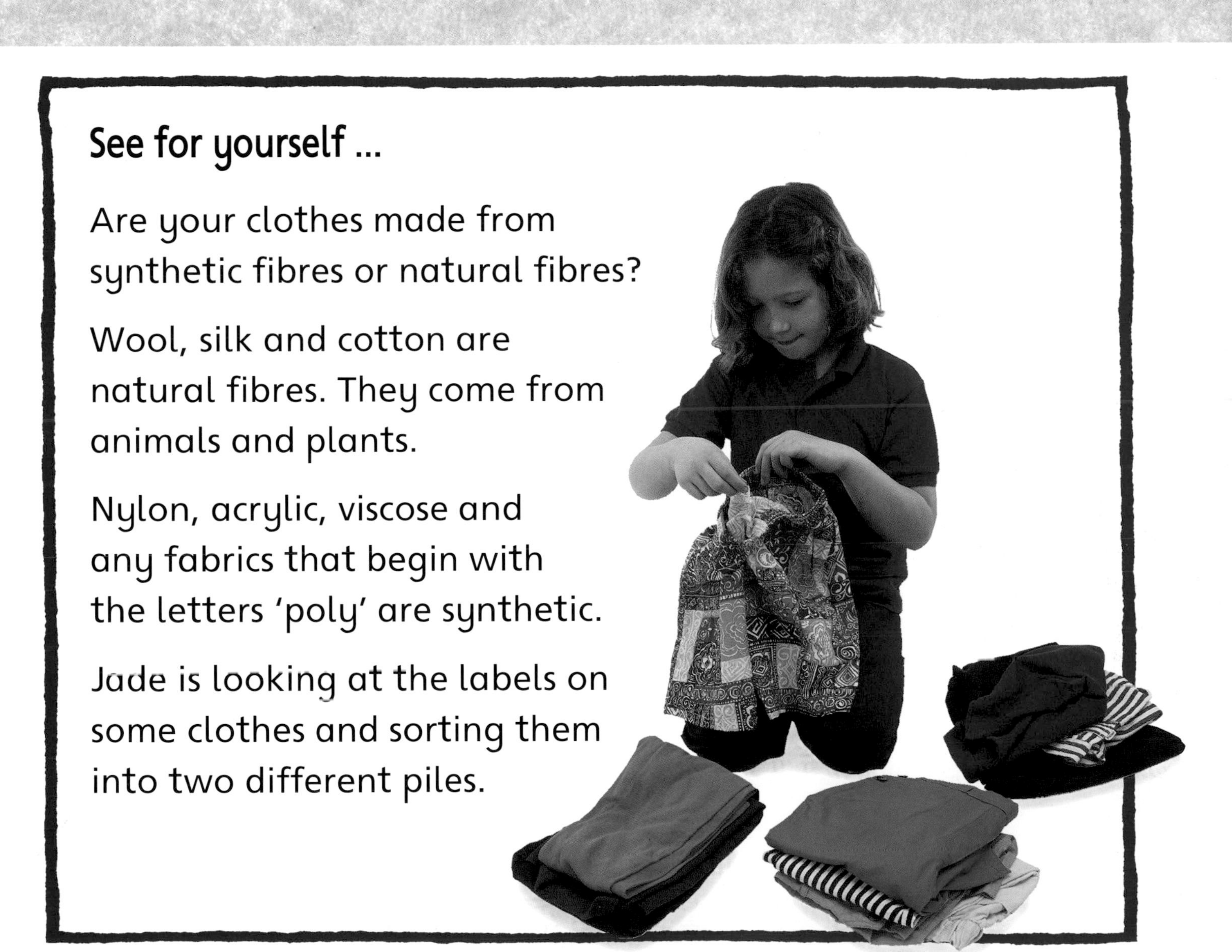

Liquids

Materials can be grouped into liquids, solids or gases.

Liquids can be poured. Some liquids, such as milk, are easy to pour.

Others, like the ketchup in this picture, are thicker and more difficult to pour.

? *How many different kinds of liquid can you think of?*

See for yourself ...

Liquids can change shape. They take the shape of the **container** they are in.

Alex has found two containers that are different shapes. She poured some juice into one container. Now she is tipping the juice into the other container.

The juice changes shape to take the shape of the second container.

Solids

Solids have a shape that usually stays the same.

They generally take up the same amount of space and cannot be squashed into a smaller space.

Most of the things that we would call objects are solids.

? *What solids can you see in this photograph?*

See for yourself ...

Robert's mum has collected together some solids and liquids for Robert to sort out.

Robert is going to make a group of solids and a group of liquids.

In which group do each of these things belong?

Gases

Gases are materials that you cannot, usually, feel.
Yet gases are all around you.
Gases have no shape and can spread everywhere.

This boy's arm bands are filled with air.
Air is a mixture of gases.

Can you think of any other things that you fill with air?

See for yourself ...

Gases can be squashed.

When James blew up the beach ball, the gases filled it out.

When he squeezes the ball, the sides go in, as the gases squash up.

When he lets go, the sides will go out as the gases spread out and fill up the space again.

Properties

The properties of a material tell you about the way the material looks, feels and **behaves**.

Materials can be wet or dry, rough or smooth.
You can see through some materials but not others.
Some materials can be squashed.

? *What words would you use to describe this tree trunk?*

See for yourself ...

Kitty is going to sort some objects into a group of hard objects and a group of soft objects.

Then she will sort them into shiny objects and **dull** objects.

After that she will sort them into rough and smooth objects.

Will the groups always contain the same objects?

Uses

The properties of a material decide how a material is used.
The material has to be right for the job it has to do.

Glass is just right for windows.
It is see-through, waterproof, rigid and easy to clean.

? *Why are windows not made from paper?*

See for yourself ...

Kitty, Sam and Alex are all wearing hats.
Each hat has been made from a different material.

Which hat is meant for a rainy day?
Which is meant for a cold day?
Which is meant for a hot, sunny day?

Heating and cooling

Materials change when they are **heated** or cooled.

Liquids will freeze if they get cold enough.
Solids will melt if they get hot enough.

The rock that is coming out of this **volcano** is so hot that it has turned into a liquid.

As the liquid rock cools down, it will turn back into a solid.

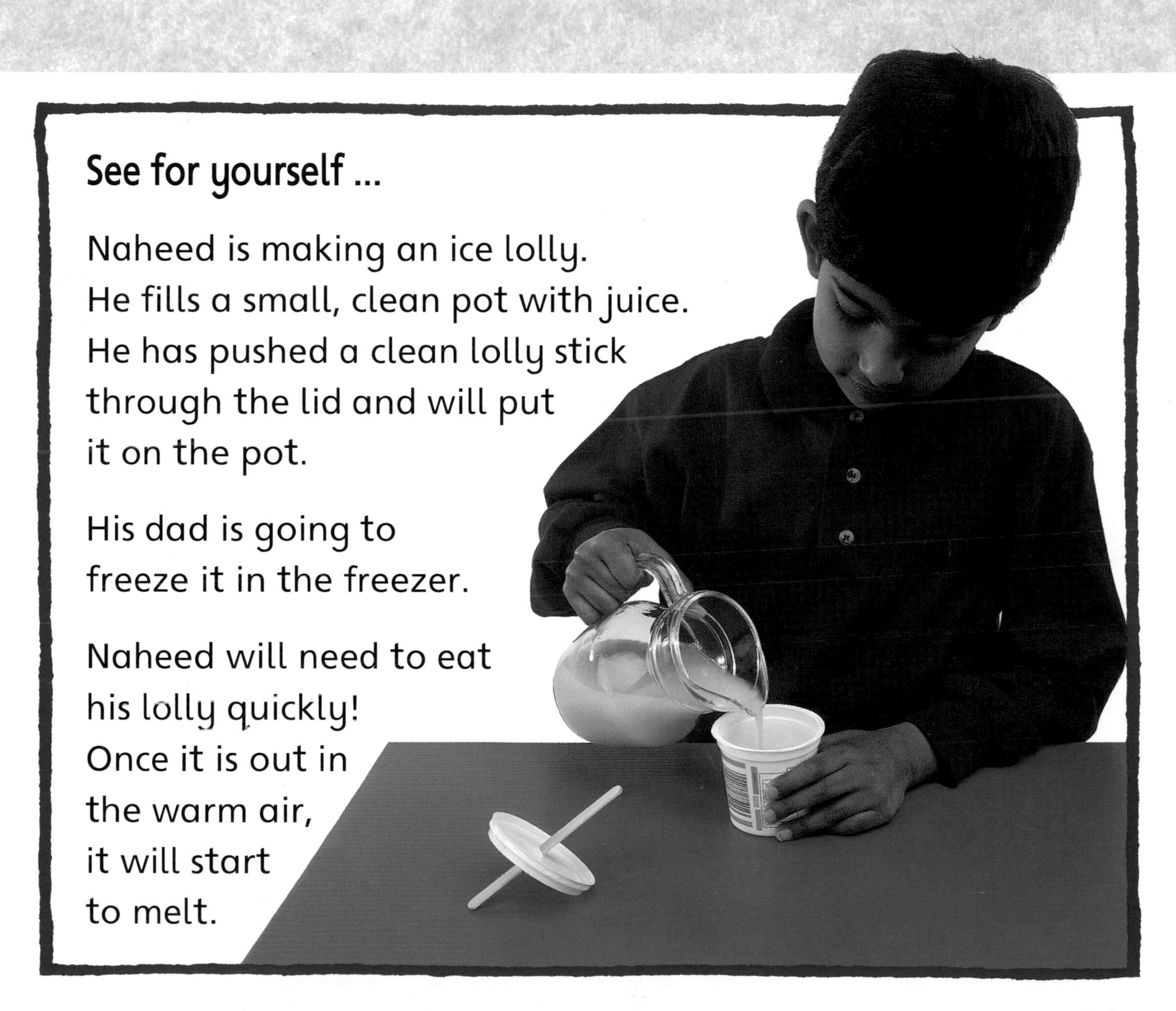

See for yourself ...

Naheed is making an ice lolly.
He fills a small, clean pot with juice.
He has pushed a clean lolly stick
through the lid and will put
it on the pot.

His dad is going to
freeze it in the freezer.

Naheed will need to eat
his lolly quickly!
Once it is out in
the warm air,
it will start
to melt.

Glossary

behaves acts

container an object that can have things inside it

dull has no shine

fibres threads

heated made hotter

synthetic anything made by mixing two or more materials

volcano a mountain with a hole in the top, through which hot rock may escape

Index